GROSS AND DISGUSTING PARASITES

A Crabtree Branches Book

Julie K. Lundgren

CRABTREE
Publishing Company
www.crabtreebooks.com

School-to-Home Support for Caregivers and Teachers

This high-interest book is designed to motivate striving students with engaging topics while building fluency, vocabulary, and an interest in reading. Here are a few questions and activities to help the reader build upon his or her comprehension skills.

Before Reading:

- *What do I think this book is about?*
- *What do I know about this topic?*
- *What do I want to learn about this topic?*
- *Why am I reading this book?*

During Reading:

- *I wonder why...*
- *I'm curious to know...*
- *How is this like something I already know?*
- *What have I learned so far?*

After Reading:

- *What was the author trying to teach me?*
- *What are some details?*
- *How did the photographs and captions help me understand more?*
- *Read the book again and look for the vocabulary words.*
- *What questions do I still have?*

Extension Activities:

- *What was your favorite part of the book? Write a paragraph on it.*
- *Draw a picture of your favorite thing you learned from the book.*

TABLE OF CONTENTS

HIDDEN HORRORS

Turn these pages to discover an unseen world of small savages. Parasites are life-forms that live inside or on another living thing. They steal food and other resources, and can cause harm, sickness, or even death.

A Frightful Parasite! Eating while reading about parasites is not recommended!

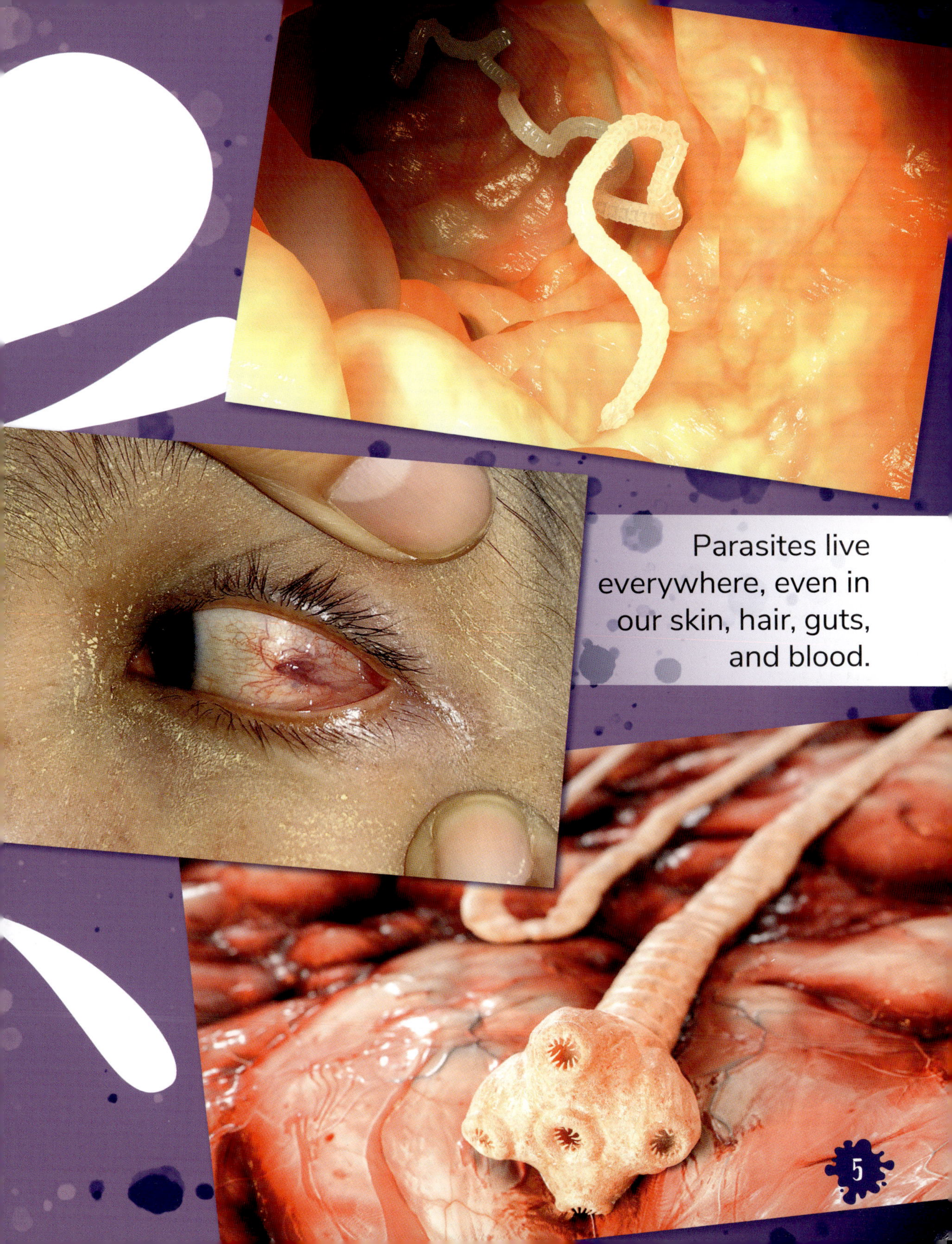

Parasites live everywhere, even in our skin, hair, guts, and blood.

Some parasites live inside their **hosts.** Others live on the outside of their hosts. Parasites can have simple or complex life cycles.

Budgerigars can get scaly face, a disease from parasitic mites.

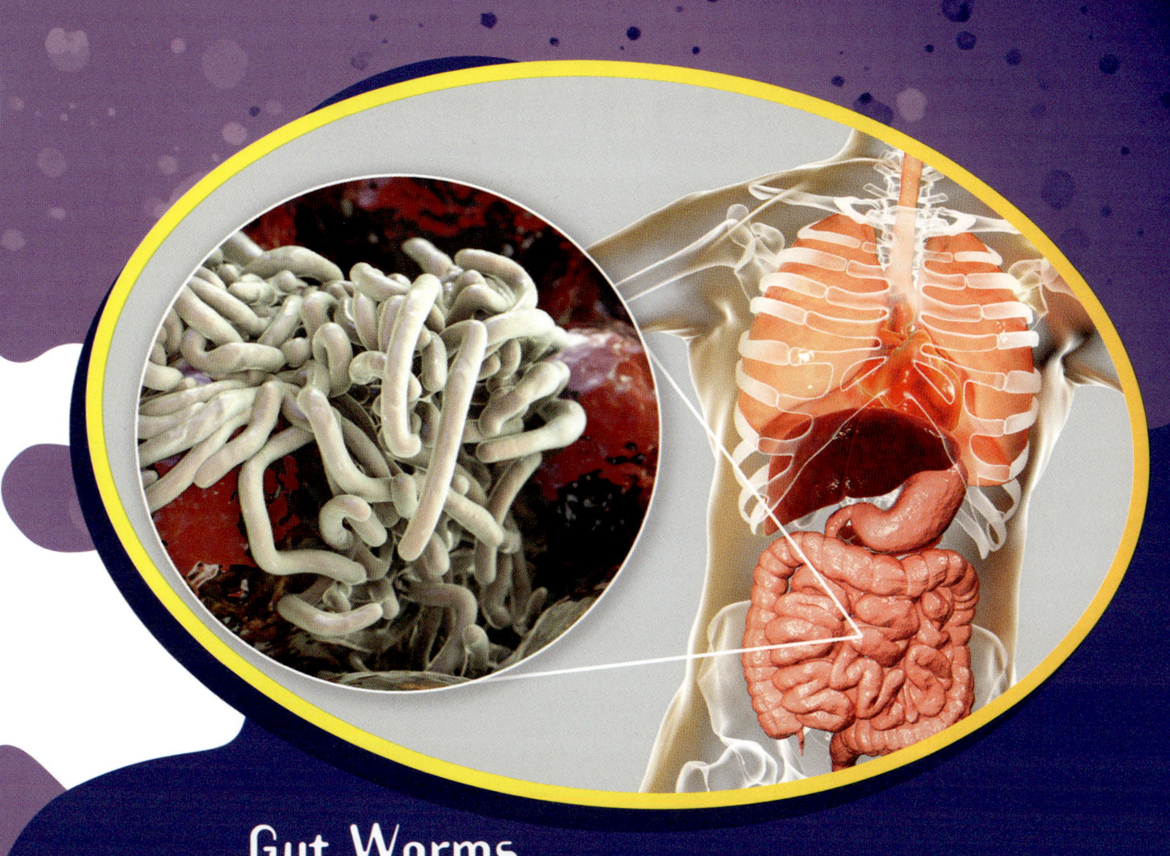

Gut Worms

Microscopic eggs in food or water may hatch and grow into long worms inside the human gut.

UNWANTED GUESTS

Parasites use host animals and people as free food, homes, and transportation.

Animals groom themselves and each other to remove parasites from fur and feathers.

Not Welcome!

Head lice snuggle on the human scalp, particularly behind ears and at the back of the neck where it's toasty warm. They drink blood, grow, and lay nits. Nits hatch in about a week. Kick out these common, itchy creeps fast with special shampoo and a fine comb.

Sometimes people get parasites from pets. Dogs and cats can carry hookworms, which are in their guts and poop. People become hosts when they come into contact with infested soil or poop.

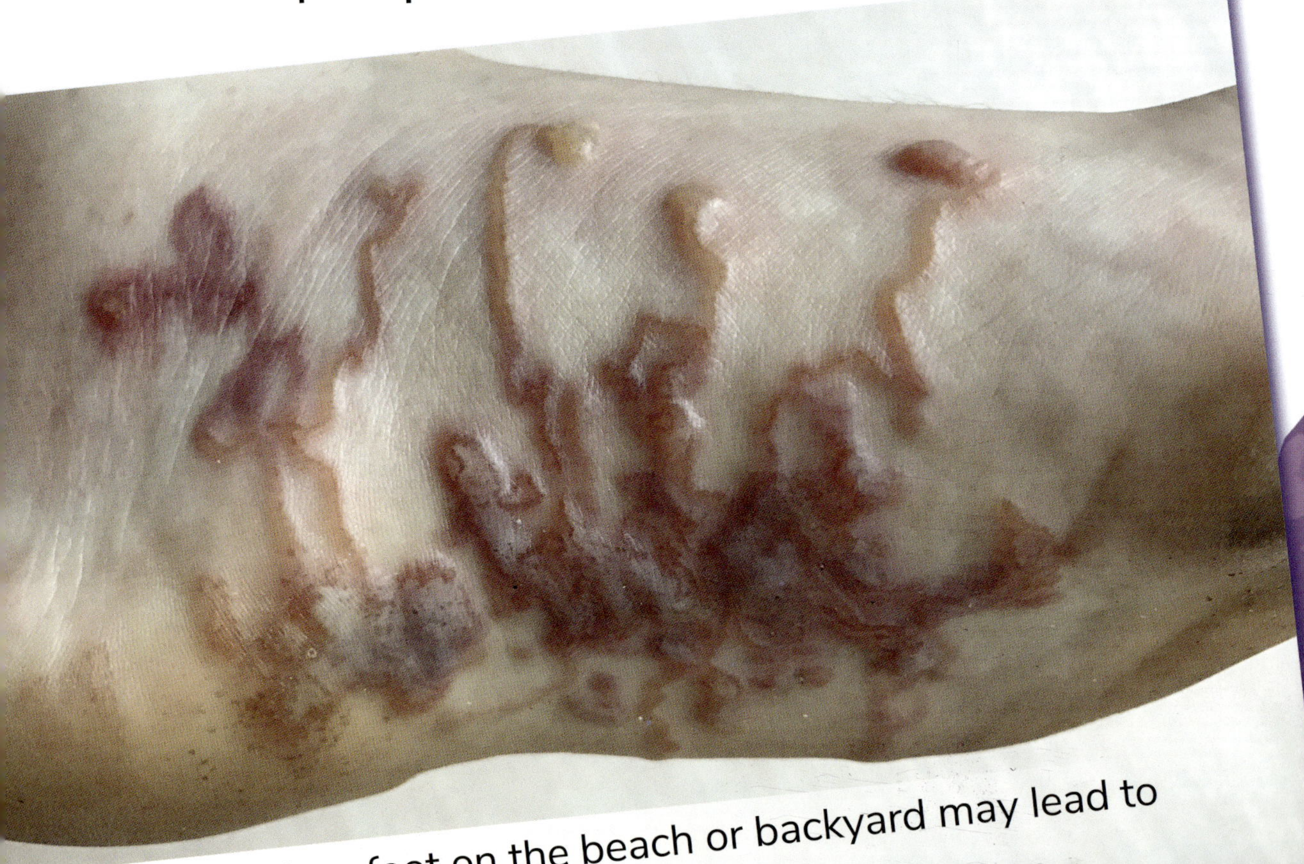

Walking barefoot on the beach or backyard may lead to a hookworm infestation.

Larva Migrans
(hookworm infection)

Mites burrow into skin to feed and lay eggs. They cause an itchy, oozing, red rash. Mites spread easily from person to person.

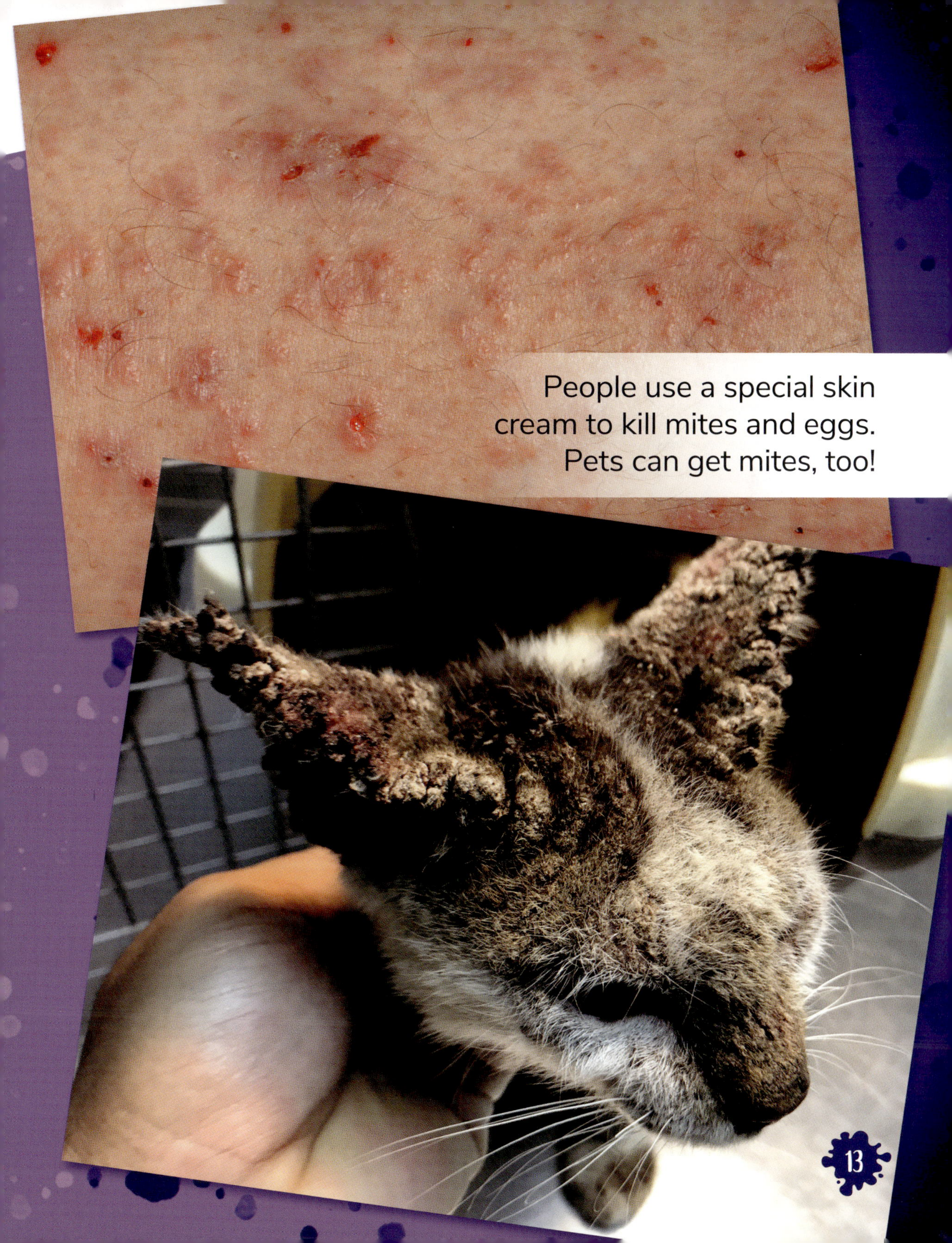

People use a special skin
cream to kill mites and eggs.
Pets can get mites, too!

13

Fish carry parasites, too. The tongue-eating louse attaches to a fish's tongue. Over time it replaces the tongue and feeds on blood and fish **mucus.**

The tongue-eating louse infests the mouths of fish we eat, like red snappers.

Creepy Cousins
Tongue-eating lice are related to crabs, shrimp, and lobsters.

Some parasites only harm certain hosts. **Giardia** live in water and soil. They infest many animals without much harm. People, however, suffer from vomiting, **diarrhea**, and gas.

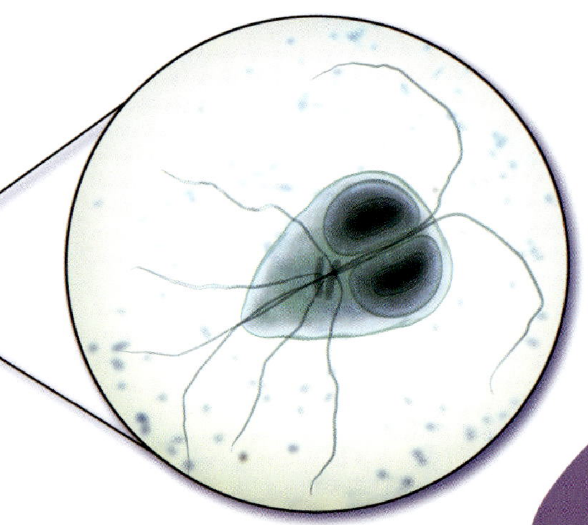

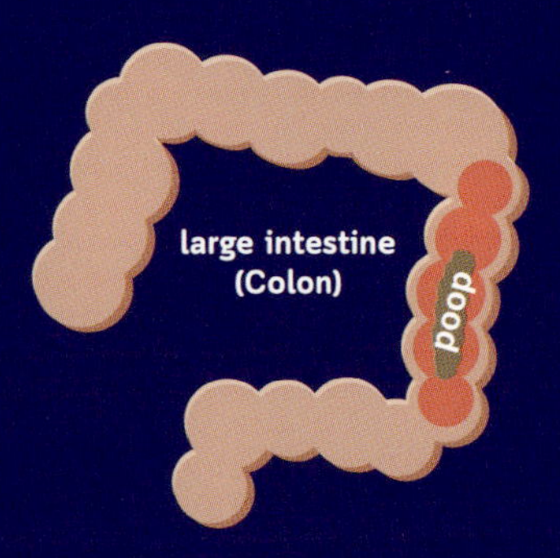

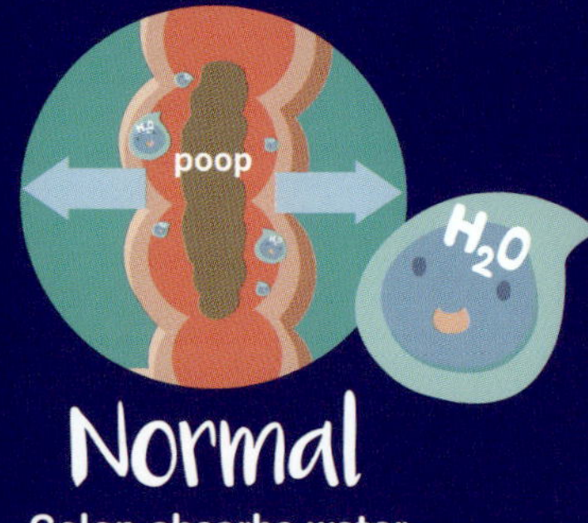

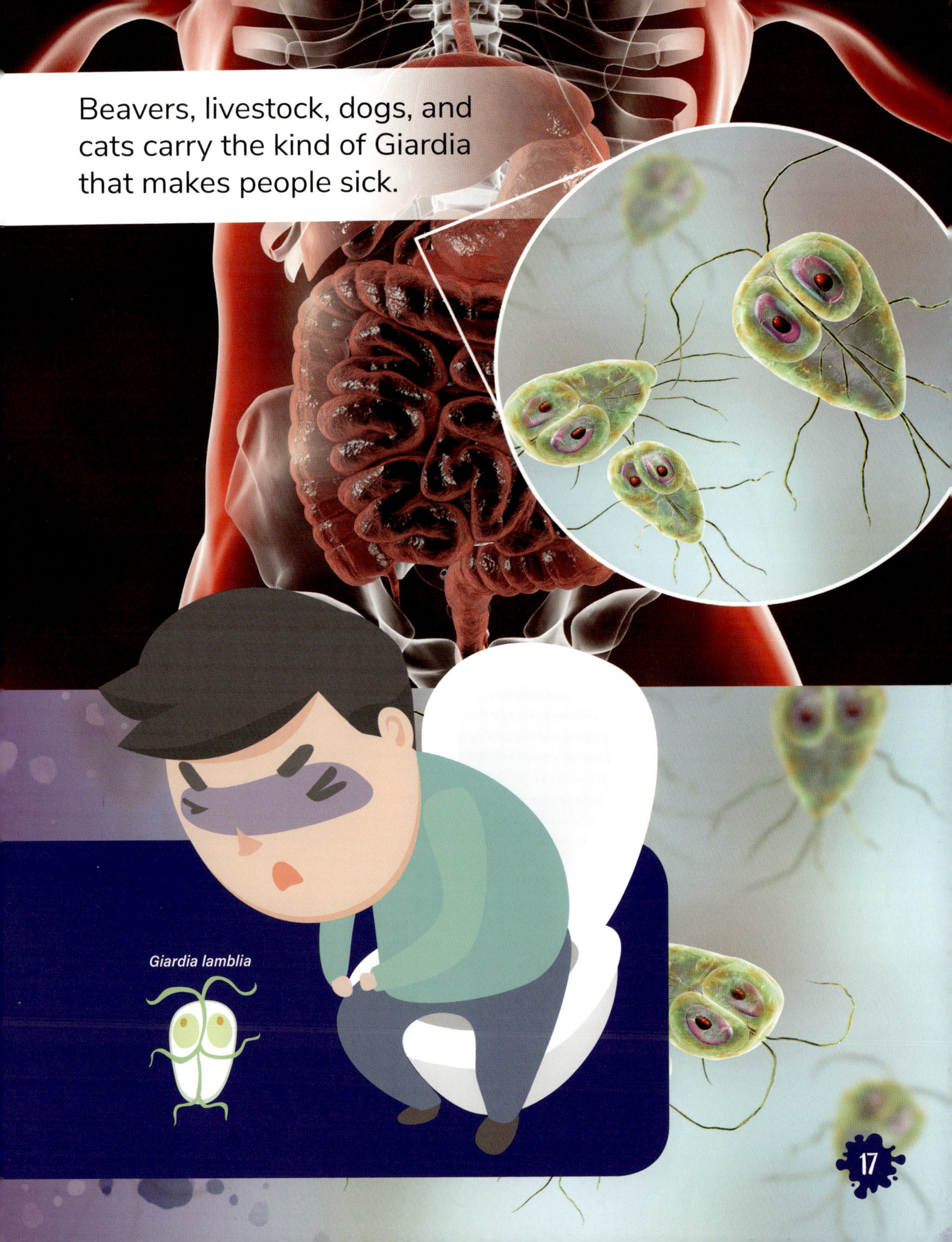

Beavers, livestock, dogs, and cats carry the kind of Giardia that makes people sick.
Giardia lamblia
17

MIND CONTROL

Did you know certain parasites can cause brain changes, turning their hosts into zombies?

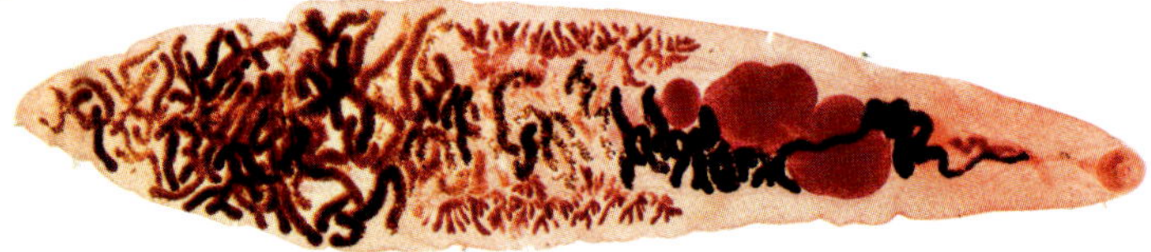

Coughing snails, living slime, zombie ants, and clueless cows are all victims of the lancet fluke.

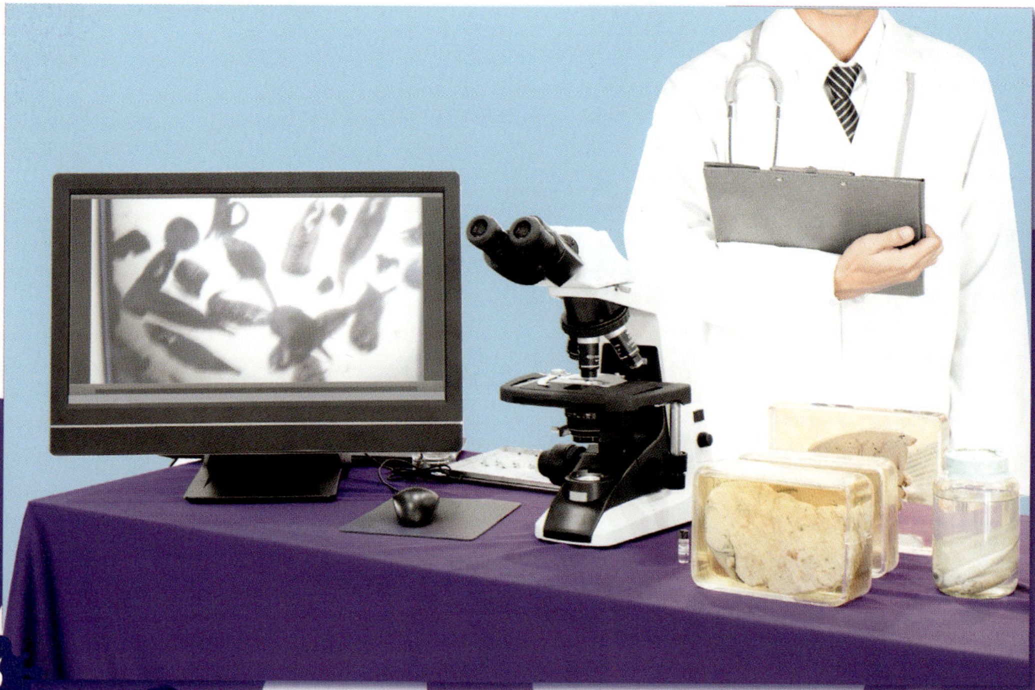

Don't Drink the Slime

Lancet flukes live in snails. Snails cough out mucus containing fluke larvae. The larvae take over the brains of thirsty ants. At night, infested ants climb up blades of grass and lay paralyzed. During the day, the ants behave normally. This repeats daily until they are eaten by the main host, cows! Cows poop out fluke eggs, which become snail snacks.

Brainworms infest their main host, white-tailed deer, when the deer accidentally eat infected snails while grazing. While not a problem for deer, this parasite is a nasty problem for moose.

Moose Madness

Brainworms tunnel in and destroy moose brains. Infected moose stop eating, have trouble walking, and die within weeks. Wildlife managers work to prevent the spread of this parasite.

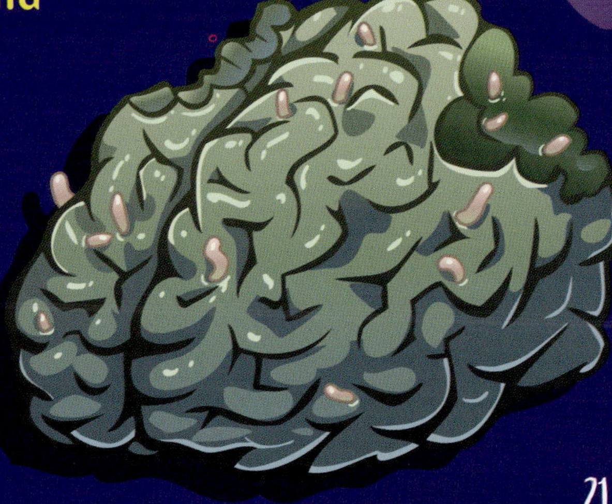

BLOOD MEALS

Many parasites feed on blood. Ticks use a pair of sharp mouthparts to dig and hook into the host's skin. The tick also uses these parts like a straw to suck blood.

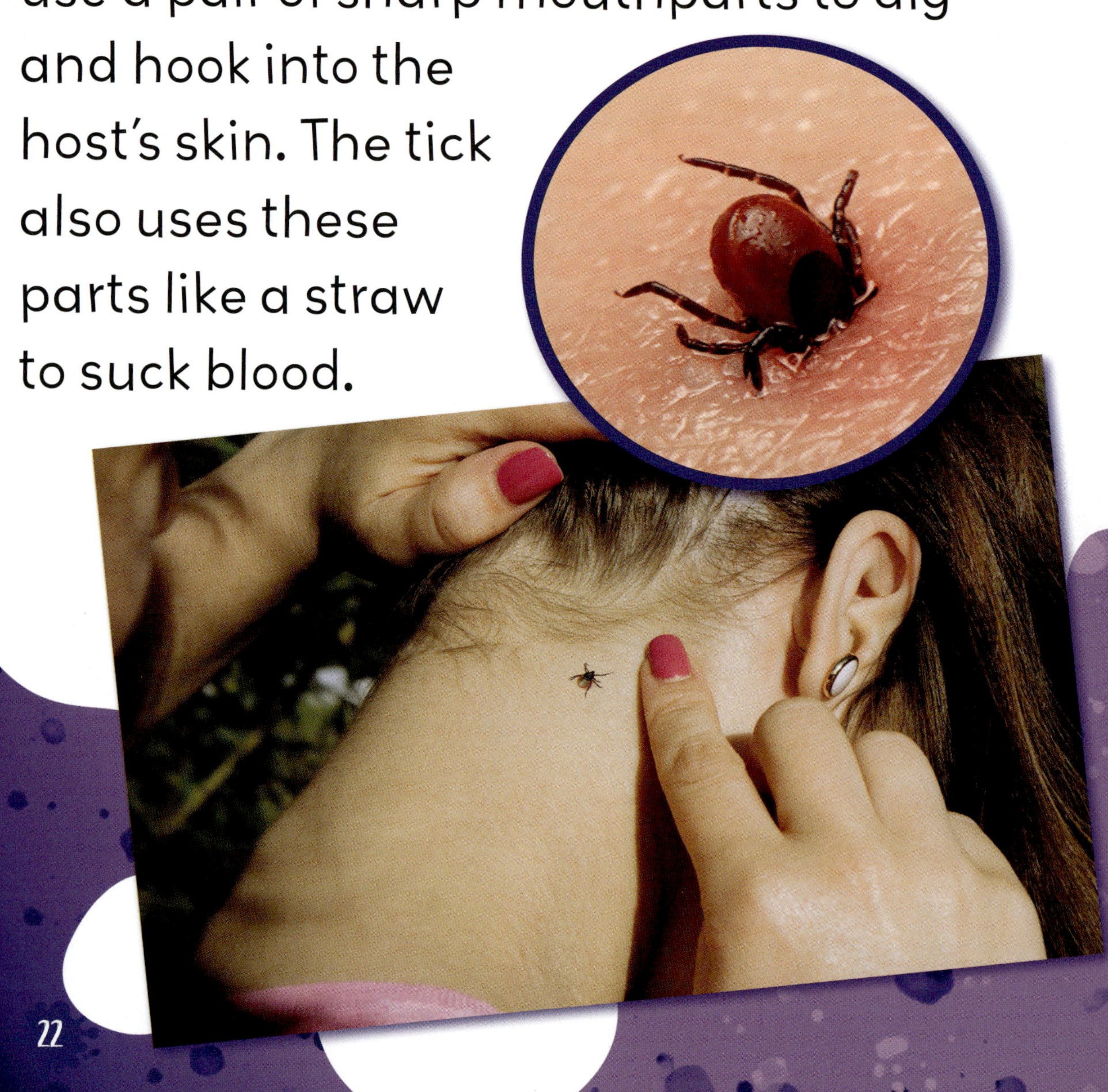

Huge numbers of feeding ticks can cause enough blood loss to weaken an animal.

More Creepy Cousins
Ticks and mites have eight legs and are related to spiders, not insects.

Many kinds of **mosquitoes** and biting flies only drink blood. Animals with tough skin still have soft spots around eyes, ears, or other areas perfect for biting.

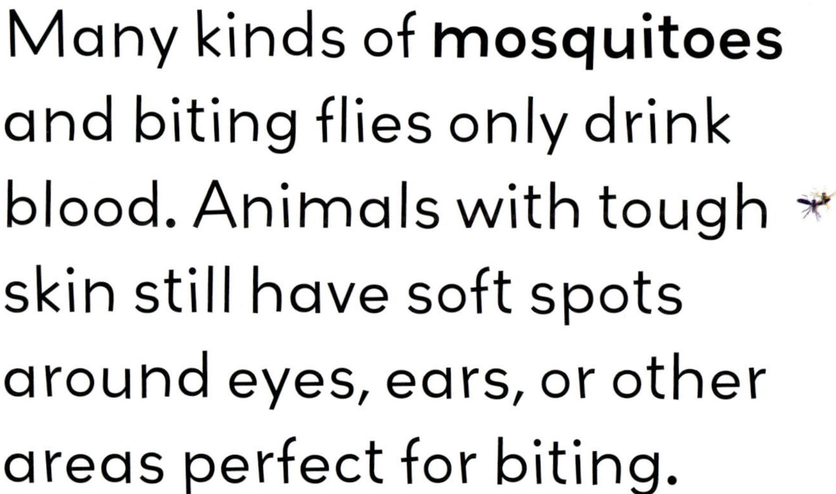

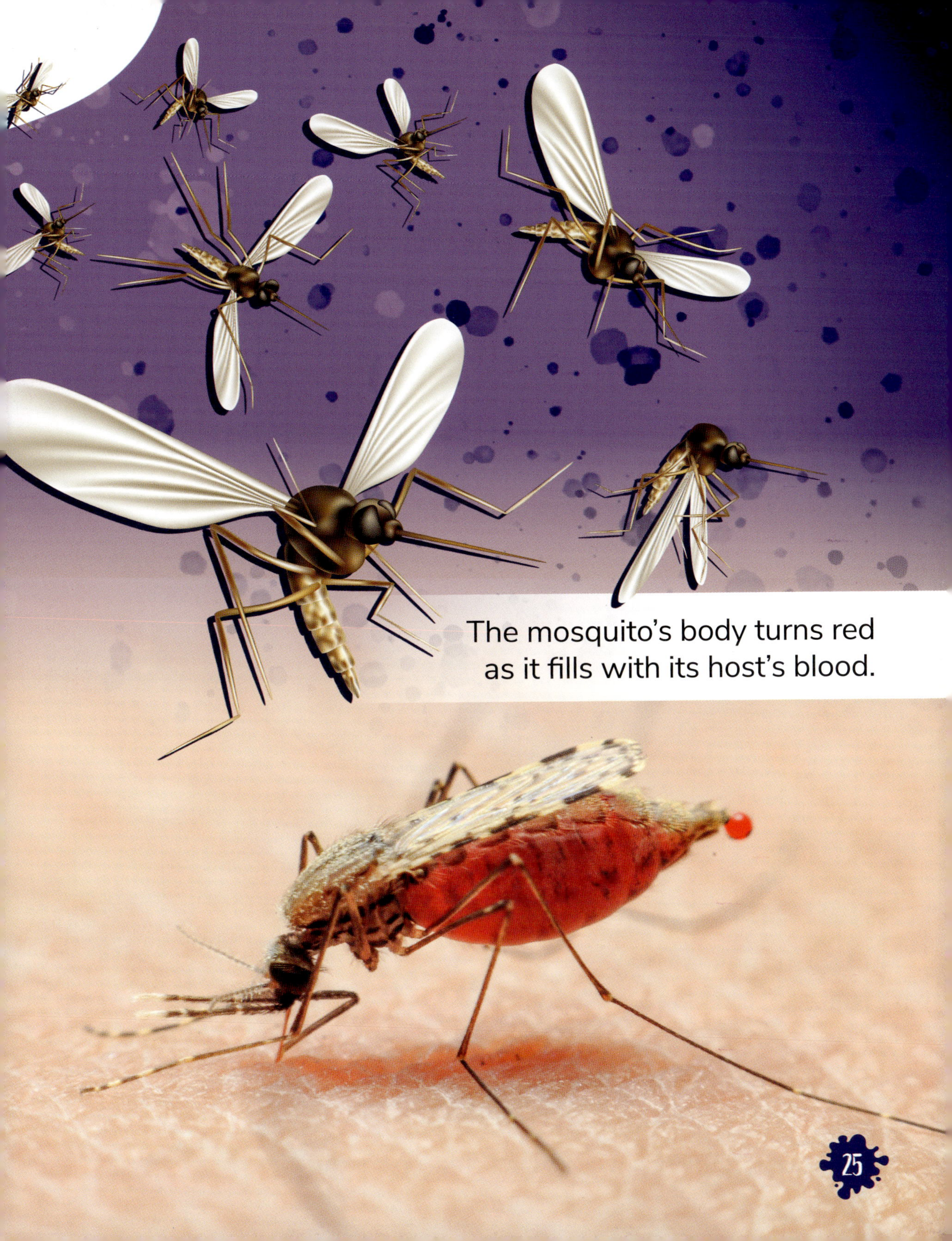

The mosquito's body turns red as it fills with its host's blood.

DISEASE CARRIERS

Parasites can infect hosts with bacteria that cause serious diseases, like **malaria.**

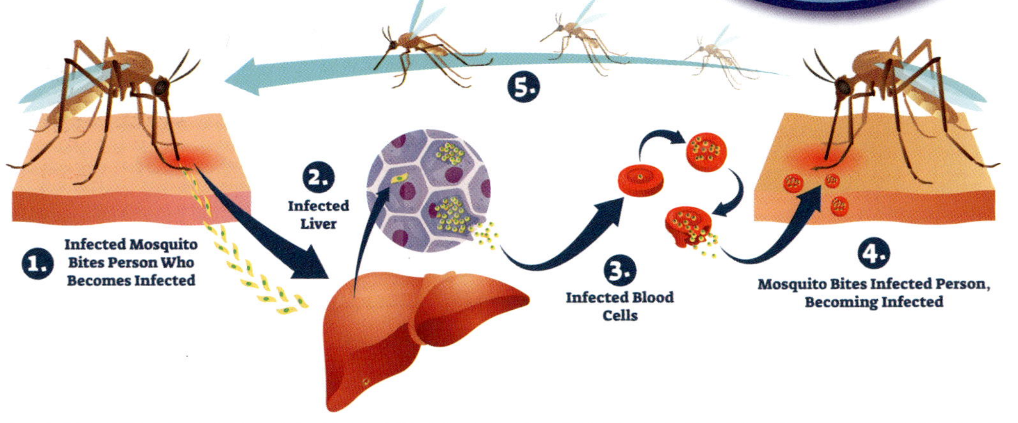

Mosquitoes cause malaria, infecting hundreds of thousands of people every year.

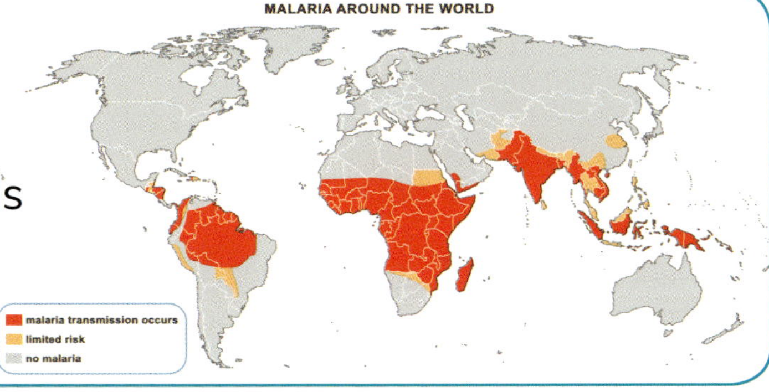

Deer ticks can infect their hosts with the **bacteria** that causes Lyme disease. The bacteria can cause weakness, fever, rash, headache, and heart and joint problems.

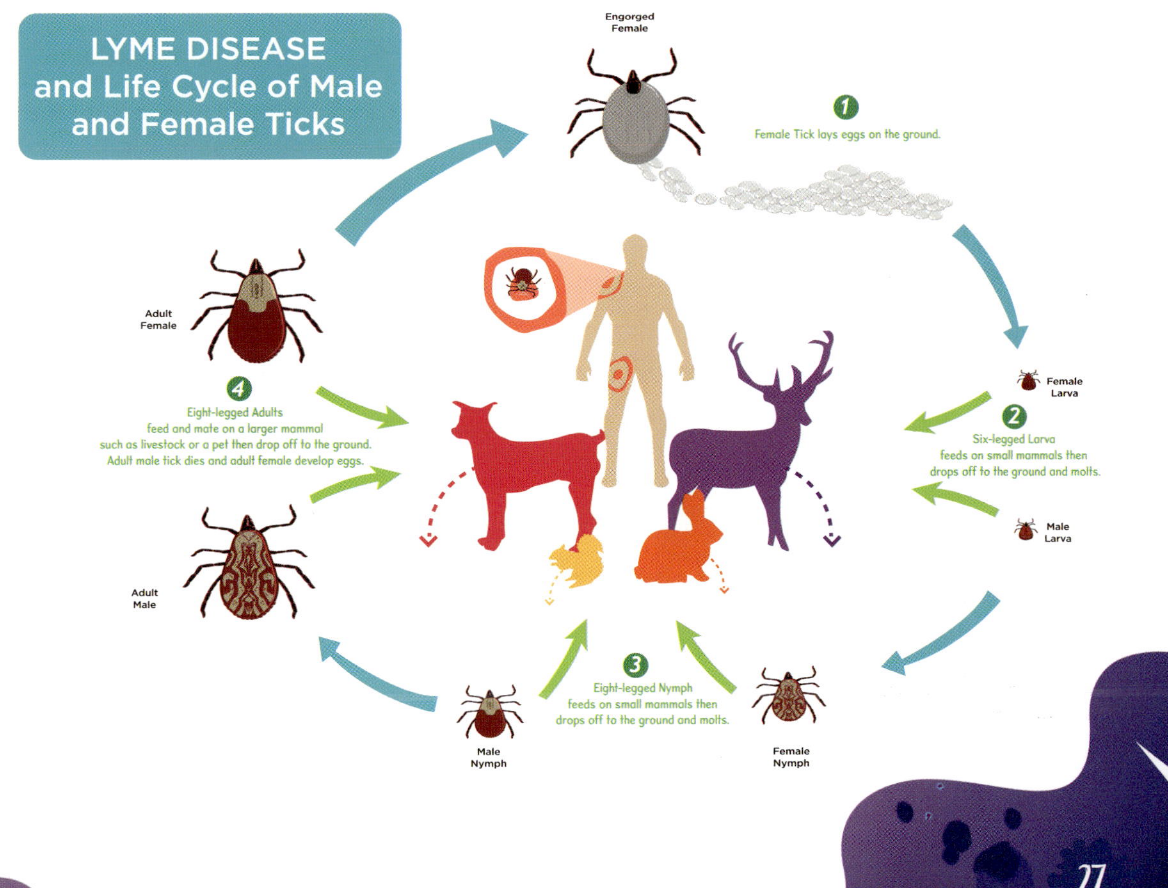

Do you find parasites gross and disgusting, yet fascinating? Parasitologists study parasites and the ways they live and act to survive. They help solve parasite problems, too. Could you?

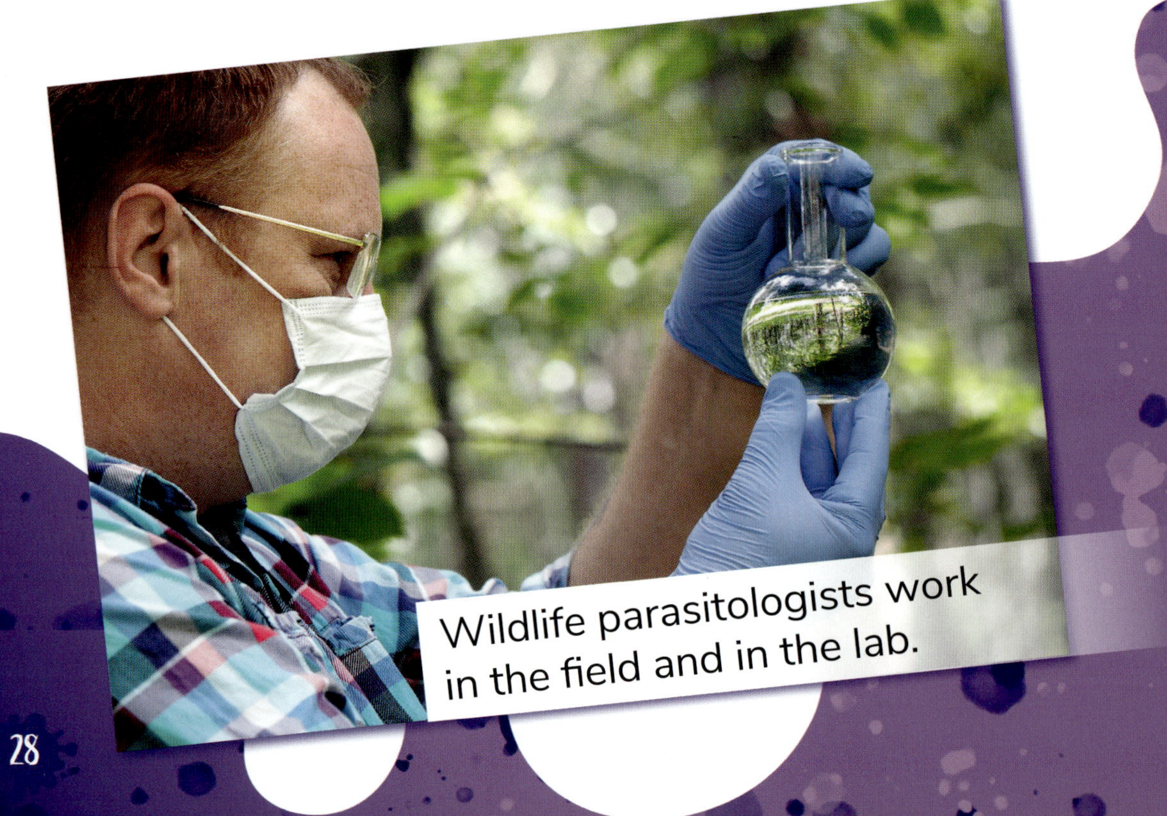

Wildlife parasitologists work in the field and in the lab.

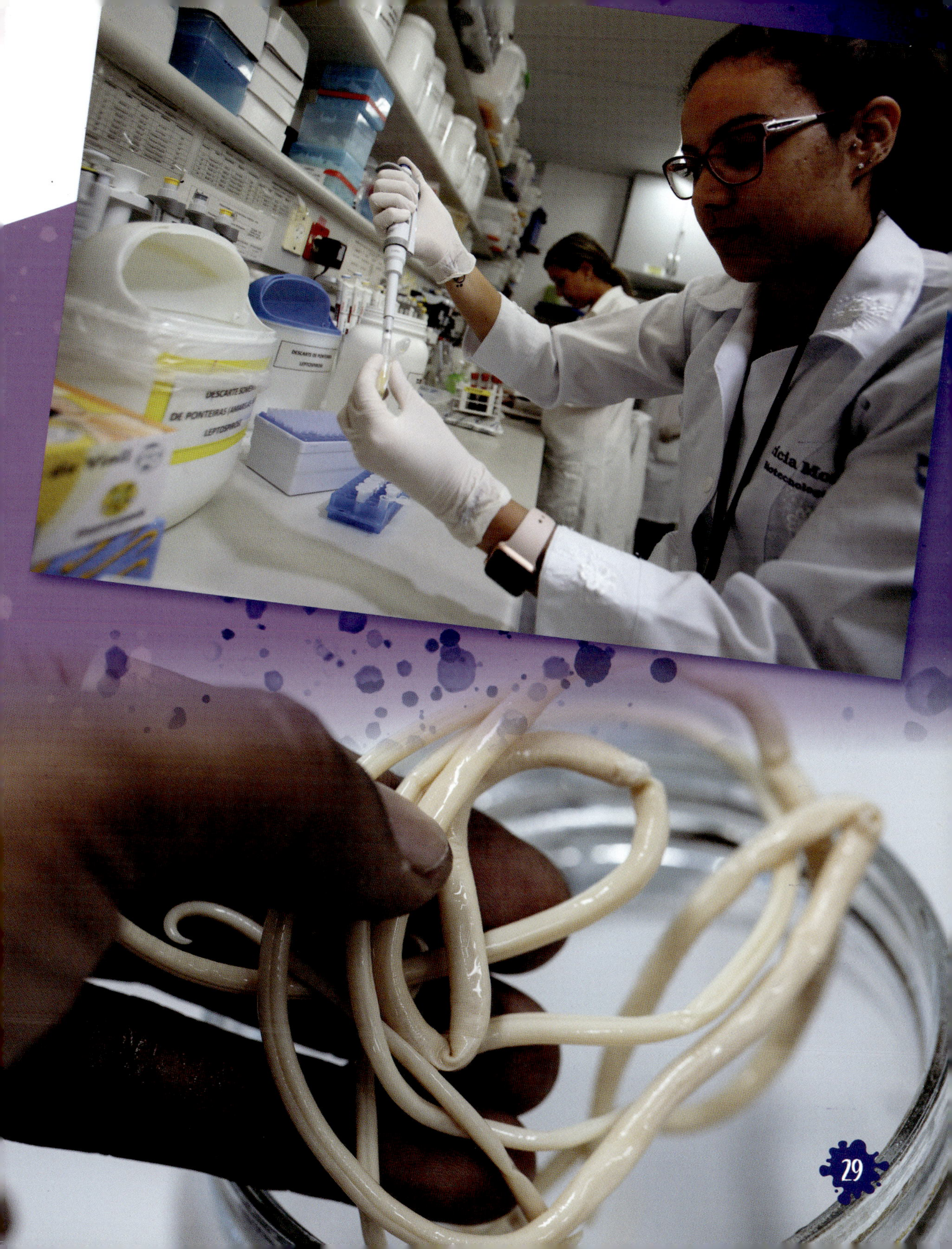

GLOSSARY

bacteria (bak-TEER-ee-uh): Common microscopic living things that can cause disease

diarrhea (dye-uh-REE-uh): Loose or liquid poop, often as a result of illness

Giardia (jee-AR-dee-uh): Common kind of parasitic bacteria found in water, soil, and animal hosts

hosts (HOHSTS): The animals on or in which parasites live

lancet flukes (LAN-sit FLUKES): Parasitic flatworms with a complex life cycle that includes snails, ants, and livestock

larvae (LAR-vee): The stage of development between egg and adult in a life cycle

malaria (muh-LAIR-ee-yuh): A tropical disease of fevers, chills, and sometimes death, caused by bacteria spread by mosquitoes

mosquitoes (muh-SKEE-doze): Flying insects with needle-sharp mouthparts for sucking blood

mucus (MEW-kuss): Spit or slime made by the body to keep areas moist and protected

nits (NITS): Eggs of insect parasites, especially lice

INDEX

WEBSITES TO VISIT

www.amnh.org/explore/ology/microbiology

https://kids.kiddle.co/Parasitism

www.petsandparasites.org/parasites-and-your-family/

ABOUT THE AUTHOR
Julie K. Lundgren

Julie K. Lundgren grew up on the north shore of Lake Superior, a place of woods, water and adventure. She loves bees, dragonflies, old trees, and science, and sometimes wonders if her cat is a parasite. Her interests led her to a degree in biology and a lifelong curiosity about wild places.

CRABTREE
Publishing Company

Produced by: Blue Door Education for Crabtree Publishing

Written by: Julie K. Lundgren

Designed by: Jennifer Dydyk

Edited by: Tracy Nelson Maurer

Proofreader: Crystal Sikkens

Cover photo © narong sutinkham, Cover splat art (on cover and throughout book) © SpicyTruffel page 4 © Vit Kovalcik, page 5 (top) © Juan Gaertner, (center) © Zay Nyi Nyi, (bottom) © Crevis, page 6 bird © Vyaseleva Elena, page 7 (top) © Kateryna Kon, (bottom) © Rattiya Thongdumhyu, page 8 © Ninelro, page 9 hair photo © khunkorn, closeup of nit © SciePro, illustration © sbego, page 10 © TisforThan, page 11 illustration at top © Designua. photo © PairutPanyamano, illustrations at bottom © StockSmartStart, page 12 both images © SciePro, page 13 (top) © Chuck Wagner, (bottom) © M. Sam, page 15 bottom photo only © Ayah Raushan, pages 16-17 all illustrations in blue box © nekoztudio, other two © Kateryna Kon, page 18 (top & in life cycle page 19) © D. Kucharski K. Kucharska, (bottom) © Suwin, page 19 snail © Violent_youth67, ants © Tuxido77, cow © WPAINTER-Std, poop © nikiteev_konstantin, page 20 © Desiree Collier, page 21 brain © Neizu, page 22 (top) © Aksenova Natalya, (bottom) © Kalcutta, page 23 (top) © Ivan Popovych, (bottom) © KPixMining, page 24 © Refluo, page 25 flying mosquitoes © Oxima, mosquito sucking blood © Aloun, page 26 petri dish © felipe caparros, mosquito infection illustration © VectorMine, map © Peteri, page 27 photo © AnastasiaKopa, illustrations © Crystal Eye Studio, page 28 (top) © Sappasit, (bottom) © Irina Kozorog, page 29 (top) Joa Souza, (bottom) © Rattiya Thongdumhyu. All images from Shutterstock.com except page 6 mite (bird parasite) © Alan R Walker https://creativecommons.org/licenses/by-sa/3.0/deed.en, pages 14-15 fish with louse in mouth and louse on spoon both images © Marco Vinci https://creativecommons.org/licenses/by-sa/3.0/deed.en, page 21 (top) © Cory Thoman | Dreamstime.com

---

Library and Archives Canada Cataloguing in Publication

Title: Gross and disgusting parasites / Julie K. Lundgren.
Names: Lundgren, Julie K., author.
Description: Series statement: Gross and disgusting things | "A Crabtree branches book". | Includes index.
Identifiers: Canadiana (print) 20210220503 | Canadiana (ebook) 20210220511 | ISBN 9781427154491 (hardcover) | ISBN 9781427154552 (softcover) | ISBN 9781427154613 (HTML) | ISBN 9781427154675 (EPUB) | ISBN 9781427154736 (read-along ebook)
Subjects: LCSH: Parasites—Juvenile literature. | LCSH: Parasites—Miscellanea—Juvenile literature. | LCSH: Parasitism—Juvenile literature. | LCSH: Parasitism—Miscellanea—Juvenile literature.

Classification: LCC QL757 .L86 2022 | DDC j591.7/857—dc23

Library of Congress Cataloging-in-Publication Data

Names: Lundgren, Julie K., author.
Title: Gross and disgusting parasites / Julie K. Lundgren.
Description: New York : Crabtree Publishing, 2022. | Series: Gross and disgusting things - a Crabtree branches book | Includes index.
Identifiers: LCCN 2021022339 (print) | LCCN 2021022340 (ebook) | ISBN 9781427154491 (hardcover) | ISBN 9781427154552 (paperback) | ISBN 9781427154613 (ebook) | ISBN 9781427154675 (epub) | ISBN 9781427154736
Subjects: LCSH: Parasites--Juvenile literature. | Parasitic diseases--Juvenile literature.
Classification: LCC QL757 .L86 2022 (print) | LCC QL757 (ebook) | DDC 578.6/5--dc23
LC record available at https://lccn.loc.gov/2021022339

LC ebook record available at https://lccn.loc.gov/2021022340

Crabtree Publishing Company

www.crabtreebooks.com   1-800-387-7650

Published in the United States
Crabtree Publishing
347 Fifth Avenue, Suite 1402-145
New York, NY, 10016

Published in Canada
Crabtree Publishing
616 Welland Ave.
St. Catharines, ON, L2M 5V6